GLOBAL WARMING, AFRICA and NATIONAL SECURITY

Global climate change is occurring. Within the last 100 years, the Earth's temperature has risen by .74°C.[1] Why it is occurring is still debated in some circles, but by the vast majority of accounts, human activity is the primary cause. What is also without contention is that the Earth's temperature will continue to rise and have adverse effects over a large portion of the globe unless greenhouse gas producers reduce their total emissions. As a result of its environmental impact, climate change can act as a threat multiplier for instability in some of the most volatile regions in the world[2] and significantly affect U.S. national security interests. This paper will address the historical context and conceptual growth of climate change as a strategic issue. Once the foundation is in place, the paper will examine whether security policy can impact and or mitigate the affects of regional climate change and more specifically what is the U.S. Combatant Commander's role in the mitigation process. Based on the significant impact climate change will have on Africa, the paper will closely look at the new AFRICOM Headquarters and how it should plan to support and mitigate environmental challenges and possible crises exacerbated by the phenomenon.

Climate Change History

The science behind climate change and even man's affect on the climate is not new. As early as the late 19th century, Swedish chemist Svante Arrhenius began to calculate how much CO_2 the Earth's atmosphere can contain before the temperature rises.[3] Arrhenius was continuing research started by the French scientist Joseph Fourier. Fourier was the first scientist to demonstrate that certain atmospheric gases enclosed Earth "like a bell jar". The CO_2 allows sunlight to pass through while absorbing infrared rays. The atmosphere is then heated in two ways: from above by the infrared rays and below as the Earth radiates heat as it cools at night.[4] In 1896, Arrhenius published his results. He estimated that given the current rate of CO_2 emissions (primarily from coal burning at that time) it would take a millennia to raise the Earth's CO_2 levels by 50 percent and that this would cause a 5-6°C rise in temperatures.[5] Current scientific thought validates his premise with some refinements based on long-term measurements. Scientists now indicate that the CO_2 levels will rise faster, having risen by 30 percent in the 20th century alone, and that the temperature increase will be closer to the 2-3°C range[6].

Long term measurement of global warming trends has occurred for various reasons during different periods of time since Arrhenius first published his work. In some instances, the measurement's initial use had no relation to climate change. During the Cold War, U.S. submarines patrolling the Arctic Circle routinely surveyed the thickness of the ice. When the military released the data in the 1990's, it provided what some researchers felt were some of the first signs of global warming. The measurements indicated an average 40 percent thinning (1.3m) of the ice from the first readings back in 1953.[7]

One of the first deliberate experiments to measure the baseline level of atmospheric CO_2 was started in the 1960s by researchers at Scripps Institution of Oceanography. The researchers took a large number of measurements over a multi-year period. Within only two years, the Scripps researchers were able to demonstrate a rise in CO_2 levels.[8] Outside a small circle of climatologists, this data did not attract much attention. At the time, computer models did not indicate a significant issue with global warming suggesting only a modest increase of 2°C[9].

Approximately 30 years ago, the field of climate change began to attract a wide variety of research scientists from wide ranging fields of study. The diversity of view points added tremendously to the collective body of knowledge and also attracted resources. One particularly beneficial field of study was the analysis of deep core ice samples which provided atmospheric data for hundreds of thousands of years. From this data, scientists could determine atmospheric composition and likely temperature ranges. In 1987, a core cut from central Antarctica showed that in the previous 400,000 years, CO_2 had dropped to 180 parts per millions (ppm) during the most extreme glacial periods and climbed as high as 280ppm in warmer times, but not once higher. In the outside air, CO_2 was measured at 350ppm, unprecedented for nearly half a million years.[10]

In 1988, in order to focus attention and research efforts, the United Nations established the Intergovernmental Panel on Climate Change (IPCC). This body grew into the foremost authority on climate change and functions as one of the primary clearinghouses for the latest research. The Panel recently published the fourth in a series of major reports on the human factors affecting climate change. Highlighting the growing importance of climate change, the IPCC shared the 2007 Nobel Peace Prize with former U.S. Vice President Al Gore.[11] When the Nobel Prize Committee announced the winners, they noted that climate change poses the potential for conflict and other security challenges throughout the world, especially for countries already in a vulnerable condition. Specific threats they addressed included mass migration, competition for resources and increased threats from weather related events.[12]

The body of science built up around global warming over the last 50 years is immense. As more data became available, the field of study attracted more resources until reaching its current peak with the IPCC report. Understanding current findings sets the stage for the examination of the security impacts of the projected climate changes.

Current Science

The science surrounding climate change and global warming was, until recently, a point of contention between many in various scientific fields. However, this is no longer the case. It is routinely accepted that the increase of certain gases in the Earth's atmosphere are driving an increase in overall global temperature. The increase of these gases is primarily due to human activity or by anthropogenic means. The most important and abundant anthropogenic gases contributing to the greenhouse affect include: carbon dioxide (CO_2), methane (CH_4), and nitrous oxide (N_2O).[13] Since the pre-industrial period starting in the mid-18th century, the level of these three gases has increased significantly in the Earth's atmosphere primarily due to man-made causes. Scientists have demonstrated this through the examination of numerous ice core samples.[14] In 2005, the 379 ppm^3 atmospheric concentration of carbon dioxide, considered the most important greenhouse gas, far exceeded the natural range over the last 650,000 years (180 to 300 ppm) as [also] determined from ice cores.[15] The concentration of CO_2 in the atmosphere appears to be climbing at an ever increasing rate. The IPCC report outlines significant rates of increase for the other anthropogenic gases as well. To summarize, the IPCC report indicates that observations over the past decades, coupled with new analytical techniques, make for an irrefutable argument that the air and oceans are indeed warming. Direct observations were recorded indicating increased glacial and snow melts, rising sea levels, and temperature readings showing a marked increase globally. The report concluded that not only is the climate warming, but it is warming at a faster rate than previously documented. For example, the period 1995-2006 included eleven of the warmest twelve years on record.[16]

The vast majority of scientists who are studying global climate change concur with the anthropogenic hypothesis, but a small minority feels there are other possible causes, and some even suggest that a significant increase in CO_2 is overall beneficial to the environment. The alternative camp's research is leading them to look at such possibilities as solar anomalies, natural climatic cycles, and inaccuracies in current climate modeling systems. Another group is not sure what the effects of global warming will be on the environment and whether it holds the possibility of negative or, in some cases, positive impacts. In most cases, the arguments made by this small group of skeptics are refutable. Two prominent global warming skeptics who fall in these groups are Marlo Lewis, a senior fellow at the Competitive Enterprise Institute and Dr. Robert M. Carter of James Cook University, Townsville, Australia. In his testimony before the Senate Committee on Environment and Public Works in December 2006, Dr. Carter indicates that increasing carbon dioxide levels in the atmosphere are "beneficial to plant growth". He asserts that carbon dioxide levels in the range of 200-1000 ppm have been shown to increase plant growth and to increase their efficiency of water use. His contention is that "CO_2 is therefore a benefice."[17] There is recent research from the University of Michigan which indicates that, over time, grasslands (which account for 30 percent of the world's arable land) will be unable to sustain growth and productivity with increases from the CO_2 enriched environment projected for mid-century. The increase in CO_2 limits the nitrogen available to the plants.[18] Obviously, there is an optimum level of natural gases in the atmosphere which would benefit both plants and humans. Doctor Carter presents his findings without fully accounting for the long-term detrimental affects of CO_2.on the biosphere.

Likewise, in a commentary refuting recent media reports concerning global warming, Mr. Lewis contends that "even though 2005 was the hottest year in the instrumental record, it falls exactly on the non-alarming 0.17°C per decade trend-line of the past 30 years."[19] His assertion is that we are on a constant upward, but "steady", temperature trend and this is "OK". His commentary would imply that as long as the upward trend is steady, but not increasing, that this is not going to have detrimental effects on the global environment. Most climatologists would not agree with this assessment. Mr. Lewis' work also downplays positive feedback loops where the effects of increased temperature, even at a steadily increasing rate, cause increasingly more impact as the rise in temperature triggers other factors which, in turn, raise the temperature more. The importance of positive feedback loops were highlighted in the recent IPCC report. One such process the report highlights is the ice-albedo feedback where melting ice reveals darker land which absorbs more of the sun's radiant energy thereby raising the surrounding atmospheric temperature resulting in more ice melt.

The intent of this author is not to discount the work of any researcher on the subject of global warming. Research on the subject is always important. Clearly we are in a cycle of global warming that will continue as anthropogenic gases ("greenhouse gases") continue to accumulate in the Earth's atmosphere. Based on these trends, it is logical to assume that impacts from climate change will intensify as the concentration of "greenhouse gases" increase. What, however, are the implications of climate change with regard to possible security threats?

Climate Change as a Threat Multiplier[20]

The U.S. government has long recognized that environmental issues are significant considerations and of major interest with regards to international relations. These considerations could take the form of responding to natural disasters, pollution control, spread of disease, weather monitoring or other areas. Only within the last few years has the idea of climate change become a factor in national security. Currently, however, the Department of Defense (DoD) has no overarching directive or policy guidance that directs DoD organizations to address the security threats of climate change or act to mitigate its effects.[21] This is not to say that the DoD has not considered the issue from a strategic security standpoint. In 2003, the DoD Office of Net Assessment contracted with two well known futurists, Peter Schwartz and Doug Randall, to examine the issue of climate change in relation to security issues.[22] This report generated discussion, but no inclusion of climate change in the 2004 *National Military Strategy*, the 2005 *National Defense Strategy*, or the 2006 *Quadrennial Defense Review*. A strategic policy document which does address climate change indirectly as a security issue is the 2006 *National Security Strategy of the United States.* The document opens the door to the issue by indicating that stabilizing global greenhouse gases should be a consideration along with economic growth; don't sacrifice the environment for an unfettered economy.[23] This is a small first step for the administration.

On the legislative side, many members of Congress appear to recognize the urgency of the issue of climate change and national security and are taking the initial steps to bring the issue to the forefront. Senator Hillary Clinton submitted an amendment to the Fiscal Year 2008 Defense Appropriations Act requiring the Department of Defense to address national security risks posed by global warming. In September 2007, the House Subcommittee on Investigations and Oversights held hearings on the National Security Implications of Climate Change. The committee asked Former Army Chief of Staff, General Gordon Sullivan (Ret), to testify at these hearings based on his recent work on the subject. General Sullivan was part of a military advisory panel working with the Center for Naval Analysis (CNA), a nonprofit think tank, on the subject of climate change and national security. In April 2007, CNA released a comprehensive report summarizing the potential national security threats from climate change. The report's military advisory committee included General Sullivan and ten additional retired senior military officers representing each military service.[24] The panel members met with some of the world's leading climate scientists, business leaders, and others to examine the issue of climate change through the lens of their military experience and strategic security.[25] The report's recommendations are quite clear. It encourages all strategic leaders to consider the problem of climate change in a more comprehensive and timely manner. To summarize, the report concludes that global climate change is occurring and that it can act as a "threat multiplier for instability in some of the most volatile regions of the world" and this, in turn, posses a national security threat for the U.S. It continues by indicating that because climate change has the potential to directly or indirectly cause instability in multiple areas of the world, it could easily overwhelm our ability to react if necessary. It is therefore critical to start responding to the issue immediately in order to mitigate the impacts of the new security challenges.[26]

Although national security strategy policy development still lags behind the climate change research, it appears 2007 will be a turning point in the process to develop a coherent strategy in the future. Several factors contributed to the continued interest in the subject of climate change and national security starting the first half of 2007. First, the superb reputation of the CNA advisory panel members coupled with the clarity of the panel's report immediately generated awareness and impetus within the government. Second, the 110th Congress shifted control of both the House and Senate to the Democrats who currently appear more concerned about the issue and in particular the impact on national security. Third, IPCC released its 4th Assessment Report at about the same time as the CNA report initiating a peak of media awareness. Fourth, mass media within the U.S. began to regularly highlight issues of global warming and climate impacts. A continued high level of interest should maintain the issue on the minds of strategists encouraging debate and formulation of policy. But, why specifically is this so important from a security standpoint? The answer derives from our values and corresponding national interests.

The U.S. core national interests are: security of the homeland; economic well-being; stable international order; and promotion of national values.[27] Each of these national interests is currently affected by climate change. For example, the United

Nations has said that water scarcity is behind the bloody wars in Sudan's Darfur region, and in Somalia drought has spawned warlords and armies.[28] Of greater significance to the strategic planner is the degree to which climate change and other related factors will impact the U.S. national interests. These will likely increase in the future as the impacts of climate change continue to mount. This can be attributed to the intensity of the weather pattern changes, interdependence of global markets (food and oil for example), world population growth, increased demands from failing nations, regional water conflicts and climate related stressors placed on the U.S. itself.

The potential threat of climate change is beginning to impact the current Army leadership. Chief of Staff of the Army, General George W. Casey, Jr. laid out the challenges ahead to our national security. He states, "As we look to the future, national security experts are virtually unanimous in predicting that the next several decades will be ones of persistent conflict where local and regional frictions, fueled by globalization and other emerging trends, are exploited by extremists to support their efforts to destroy our way of life."[29] He identifies six specific global trends likely to fuel the potential for conflict (Globalization, Population Growth, Resource Demand, Climate Change and Natural Disasters, Proliferation of Weapons of Mass Destruction, and Failed or Failing States).[30] General Casey regards climate change as a leading potential cause for conflict. Similar to the CNA report, General Casey indicates that climate change will exacerbate pre-existing conditions to push vulnerable societies beyond their ability to cope. This will include destabilizing populations, increasing the threat of disease epidemics and setting the conditions for natural disasters to cause inordinate damage and destruction.[31]

Additionally, climate change will exacerbate the other trends as well, thereby making it an underlying root cause which will require focused attention and strategy development. The threat from climate change increases exponentially as it will likely impact multiple locations concurrently in a variety of ways. This is unlike most current one dimensional threats. Droughts or floods will cause crop loss and starvation; populations will be forced to migrate; weakened societies will come under ever increasing pressures; conflicts will start or escalate; and failing governments will either react harshly or collapse completely.[32] This scenario will not come to fruition in every climate stressed region. However, the possibility exists and the U.S. must recognize and plan for the security implications.

One can argue, as has been done here, that climate change is important to national security. It may be instructive to explore the linkage between climate change and the various levels of national security in greater detail. Dividing the range of impacts may also aid in conceptualizing the issue. At the macro-climatic or *global level*, climate change will affect rain and soil moisture levels and the Earth's surface and atmospheric temperatures. This will result in a reduction of the vital natural resources upon which mankind depends. Additionally, climate change may result in an increase in the intensity of storms, prolong droughts and increase the possibility of severe flooding. At a *geopolitical level*, deforestation, reduction of icecaps, rising sea levels and loss of habitable space will create security challenges and complicate the ability of military planners to project power, influence regional events and secure forward basing. At the *regional level*, changes in climate will challenge the stability and existence of failing states, open the door for extremist ideology and insurgencies, possibly interrupt access to vital resources, and generate instability that threatens U.S. national security interests.[33]

A stratified classification of security concerns implies a broad range of issues which requires a top-down, interagency approach to the issue. However, the sheer diversity and range of issues may create a problem too immense, complicated and costly to formulate into a coherent national-level strategic security policy which includes the issue of climate change. This appears especially true as the President and DoD focus on the more evident threats posed by the Global War on Terrorism (GWOT). Terrorism and the threat posed by weapons of mass destruction (WMD) are recurring themes running through the latest *U.S. National Security Strategy (2006) and The National Defense Strategy (2005)*. Additionally, there are separate documents published on these subjects to focus strategy formulation. They include the *National Strategy for Combating Terrorism* (SEP 2006), the *National Strategy to Combat Weapons of Mass Destruction* (DEC 2002) and the *National Strategy for Homeland Security* (OCT 2007). Is it time to elevate climate change or global warming to the same level by producing a national strategy to focus efforts in this area? It might be time. As this paper later indicates, the combatant commanders are moving ahead on some of the climate change issues based on command initiatives and the interpretation of general guidance from other policy documents. The other documents include the 2006 *Quadrennial Defense Review* (QDR) and the DoD Directive 3000.05, *Military Support for Stability, Security, Transition, and Reconstruction Operations*. The QDR addresses "preventive actions so problems do not become crises"[34] and *Directive 3000.05* indicates the immediate goal of stability operations, "is to provide the local populace with security, restore essential services, and meet humanitarian needs."[35] Commands are interpreting this guidance and are initiating and expanding efforts to anticipate and mitigate the effects of gradual climate change even though the documents do not specifically address the issue.

Gradual climate change and the corresponding impacts are challenging enough to the strategic level authorities; what if the Earth's climate abruptly changed? The DoD did examine this possibility with a 2003 study ordered by the Office of Net Assessment. As DoD did not publish any known strategic-level climate change guidance based on this report, it is unclear what actions were taken, if any, from the report. For the concerns of this paper, it is suffice to indicate that the possibility of rapid climate change is possible. The key for strategic planners is to review the appropriate research on the subject and include the possibility of rapid climate change in planning guidance in order that subordinate levels of command can act in the appropriate and timely manner, if necessary.

Climate Change and Africa

"The African continent is the most vulnerable in the world," according to Dr. Robert Watson, World Bank Chief Scientist and former Chairman of the United Nation's IPCC.[36] The predicted effects of climate change over the coming decades include extreme weather events, drought, flooding, sea level rise, retreating glaciers, habitat shifts, and the increased spread of life-threatening disease.[37] Africa has already experienced all of these events and will most likely experience each in greater intensity in the future. Reductions in soil moisture and further loss of arable land may be the most significant of the projected impacts of climate change in Africa. Extreme weather events are also likely to increase. These changes will lead to reduced supplies of potable water and food production in many areas. Such changes will add significantly to existing tensions and can weaken governance, cause economic collapse, massive human migrations, and potential conflicts. In Somalia, alternating droughts and floods led to migrations of varying size and speed and prolonged the instability on which warlords thrived.[38] Ironically, although Africa contributed the smallest amount of anthropogenic emissions of greenhouse gases (between 1.9 percent and 3.1 percent of global emissions between 1973 and 2002), it will likely suffer the most from the affects of climate change. This is due primarily to existing underlying issues and the lack of adequate resources to address issues as they develop.[39] In order to frame the issue and understand the depth of the climate change problem on the continent, it is important to understand some of the specific issues in detail. The areas of water, food, disease, agriculture and population growth deserve closer examination.

Water

About 25 percent of Africa's population - nearly 200 million people - do not have easy access to water; experts expect that figure to climb by another 50 million by 2020 and more than double by the 2050s.[40] According to the IPCC, of the 19 countries in the world classified as water stressed, most are in Africa.[41] In Africa, as in most places, water is at the heart of most of climate change issues. Most people would naturally assume that drought would result from the increase in global warming and the resulting effects on the Earth's climate. They can envision the mass migrations and starvation due to crop loss, deforestation, and lack of drinking water. However, this is only one side of the two-sided water coin. At times, parts of Africa will have too much water, too quickly. Stephen Schneider, a lead author for the IPCC report, summarizes the problem, "As the air gets warmer, there will be more water [vapor] in the atmosphere. That's settled science. But where, and when, it comes down is the big uncertainty. You are going to intensify the hydrologic cycle. Where the atmosphere is configured to have high pressure and droughts, global warming will mean long, dry periods. Where the atmosphere is configured to be wet, you will get more rain, more gully washers [intense downpours]. Global warming will intensify drought, and it will intensify floods."[42] Recent African news media anecdotally reports both an increase in drought and flash floods within the past several years.

Shared water resources hold the potential for future conflict. Large river basins, shared between several countries, account for approximately 85 percent of Africa's water resources.[43] Although rare in recent history, one can reasonably assume that inter-state conflict over water will likely increase in the future. A few countries in Africa have already exceeded their industrial requirements for water. This condition limits their future economic growth. Even South Africa, the strongest economy on the continent, faces the issue of water shortages impacting economic development. The trend in Africa is that more countries will also reach this point before 2025.[44] This condition sets the stage for possible regional conflicts over scarce water supplies.

Disease

Scientists working with the World Health Organization (WHO) are researching the linkage between global warming and precipitation trends and their impacts on mortality. Initial research indicates that approximately 150,000 humans die annually from direct and indirect impacts of climate change. The causes include the increased spread of disease, malnutrition, floods and heat related deaths to name a few.[45] Chronic disease is unfortunately a situation a large portion of Africa is very familiar with. Malaria and other infectious diseases routinely account for the majority of deaths in sub-Saharan Africa.[46] The 2001 IPCC report indicates vector-borne diseases, worsening pollution, dangerous water, and collapsing urban infrastructure as populations move into urban areas, as all potential consequences of climate change.[47] The opportunity for the spread of infectious diseases in densely packed urban areas will likely increase as disruptive weather patterns force populations from rural areas to urban. The infrastructure associated with sanitation systems is expensive.[48] Adding new systems to meet the challenges of a growing demand is difficult. Additionally, increased temperatures of coastal waters could also aggravate cholera epidemics in coastal areas.[49] All these factors together will require concerted effort by stable African governments to prevent societal breakdowns. The question is whether these governmental systems can further develop and remain functioning in light of the challenges.

Systems supporting the health of a population are often the first disrupted during periods of conflict or natural disasters. Referred to as water, sanitation and hygiene (WASH) systems, the system breakdown is one of the primary causes of the spread of infectious disease.[50] These systems are already stressed in many areas in Africa. The HIV epidemic is a health issue already challenging many African nations. The WHO summarized the grim facts in December 2006. "Almost two thirds (63 percent) of all people living with HIV globally live in sub-Saharan Africa, an estimated 24.7 million in 2006. Some 2.8 million more adults and children became infected with HIV in 2006, more than in all other regions of the world combined. The 2.1 million AIDS-related deaths in sub-Saharan Africa represent 72 percent of global AIDS deaths."[51] Climate change aside, this issue alone stresses the health and governance systems of the region through the untimely deaths of citizens in social and economic leadership positions. These include, teachers, technicians, military members, police officers and other professionals. Loss of people from these professions leaves a critical void with the possibility of producing millions of under-educated and under-supervised young adults.[52] The added burden and disruption caused by climate change may possibly cause many societal structures to collapse thus contributing to further instability, causing friction between groups and possibly failed states. The result is climate change fulfilling the function as the CNA report indicates – "climate change as a threat multiplier."[53]

Agriculture

William Cline, author of *Global Warming and Agriculture*, studied the affects of climate change on Africa closely and his research foresees significant agricultural impacts for the continent. Temperatures in developing countries, which are predominantly located in lower latitudes, are already closer to, or beyond, thresholds at which further warming will reduce rather than increase agricultural capacity.[54] Cline emphasizes that agriculture accounts for a much larger share of GDP in developing countries than in industrial countries and therefore disruptions to this sector will have a greater adverse impact than on more industrialized countries. (The World Bank estimates agriculture accounts for approximately 30 percent of sub-Saharan Africa's GDP[55] and 55 percent of the total value of its exports[56]). Overall, Cline predicts a 17-28 percent decline in agriculture production for Africa over the coming decades.[57] One of the reasons Africa's agriculture will not be able to adapt as other regions might is due to the lack of irrigation systems. Where other parts of the world use irrigation if necessary to support agriculture, as a continent, Africa is much more dependent on rainfall to support its agriculture and therefore much more reliant on the weather to ensure stable production. Approximately 95 percent of all African agriculture is rain fed.[58] The changing rainfall pattern affects agriculture and reduces food security, worsens water availability, and increases weather related disasters. Cumulatively, these conditions undermine economic growth and cause instability and possibly conflict.[59]

Population Factors

A high percentage of the poorest countries in the world are in Africa. Many of these same countries have the highest rates of population growth and also the youngest populations in the world. The current population of Africa is approximately 800 million people, about evenly divided between Muslims and Christians.[60] The age of a population can have a direct correlation to civil unrest. Researchers have collected data which indicates that a society which has a youth bulge (classified as a high percentage of the population between the ages of 15-29 years old) is at a higher risk for civil strife and possibly conflict. Africa has a number of countries which meet the youth bulge criteria. Although the research indicates that the youth bulge may not be the "overt" cause for conflict, it does provide a pool of young, unemployed males available for recruitment into gangs and insurgencies thereby significantly increasing the likelihood for armed conflict.[61] Other studies indicate similar trends and connections between young societies and conflict in recent decades. Between 1970 and 1999, 80 percent of the world's civil conflicts occurred in countries where 60 percent or more of the population was under the age of thirty.[62] As an example, Nigeria has a large youth bulge and a growing list of insurgent groups to include the Emancipation of the Niger Delta, or MEND.[63] The MEND membership is attracted from the large population of unemployed male youth disenfranchised by its government and the large oil companies. This sets the classic stage for a growing insurgency.

Couple the youth bulge and poverty (nearly 40 percent of Africans live below the property line[64]) with increasing stressors from climate change, and the conditions are set for a breakdown of civil order and good governance. The lack of effective governance will exacerbate the problems and likely create a positive feedback system multiplying the overall damage to fragile African nations. Unfortunately, the trend is already discouraging as Africa is the only region of the world where experts project poverty will rise within this century.[65]

Africa's Strategic Importance

This paper initially presented the background behind climate change. It is occurring and researchers are documenting new affects daily. The second portion demonstrated the detrimental impact climate change is having on Africa currently and additional areas of concern for the future. The negative impacts are significant and growing steadily. Next, the author argues why Africa is of growing strategic importance to the U.S. and what the U.S. can do to effectively engage and mitigate the effects of climate change.

The strategic importance of Africa is directly linked to the U.S. core national interests: homeland security; economic well-being; stable international order; and promotion of national values[66]. The September 2002 U.S. *National Security Strategy*

(NSS) is more specific when it indicates, "In Africa, promise and opportunity sit side by side with disease, war, and desperate poverty. This threatens both a core value of the United States - preserving human dignity - and our strategic priority - combating global terror.[67] Looking closer at the national objectives outlined in the NSS, Africa is both explicitly and implicitly discussed numerous times. In the explicit category, it mentions African issues in regards to: ending tyranny (Zimbabwe); working with others to defuse regional conflicts (Darfur, Uganda, Ethiopia/Eritrea); encouraging and supporting free trade ("African Growth and Opportunity Act"); pursuit of a free trade agreement with the countries of the Southern African Customs Union (Botswana, Lesotho, Namibia, South Africa, and Swaziland); expanding the circle of development (Millennium Challenge Corporation, President's Emergency Plan for Aids, aid to farmers and food relief, fighting corruption); and strengthening fragile and failing states (partnership with Africans to improve governance, reduce corruption, market reforms and strengthen regional bodies such as the African Union). Under the implicit category, Africa fits in the following categories: promoting effective democracies; strengthening alliances to defeat global terrorism; preventing the spread of WMD; enhancing energy security (foster conditions for increased private investment to meet growing world demand); reducing barriers to free trade (protectionism, poor governance, diminished rule of law); diversifying oil production areas; and expanding the circle of development (debt relief for heavily indebted nations, transformational diplomacy). Except for several programs, most of these goals are rather broad and non-specific. To improve the understanding of the issue, it is important to add some specificity to the NSS statements.

The globalization of world markets is a double-edged sword. On the one hand it opens more markets to consumers and on the other hand it makes these same markets dependent on each other. A small ripple in the world economic system can have world-wide impacts. The world oil market is a case in point. Minor perceived or actual disruptions in the flow of oil can cause devastation or exaggerated speculation on the world markets and then subsequently the economy. One of the reasons the U.S. continues to focus on the protection of Middle East oil is not that we get more oil from there than anywhere else, but because the world gets a high percentage of its oil from there and disruption of the flow would be devastating to the world economy. The same is now also true for oil coming from Africa.

The U.S. relies on West Africa for approximately 22 percent of its oil imports, and in the near future the total will likely be more than 25 percent.[68] Doctor Wafula Okumu provided his perspective of the U.S. vis-à-vis African oil when he testified before the House Committee on Foreign Affairs, Subcommittee on African and Global Health in August, 2007. He said, "Nigeria, Africa's largest oil producer, has now overtaken Saudi Arabia as the third largest oil exporter to the U.S. The importance of the African oil source can be gleaned from the fact that in 2006, the U.S. imported 22 percent of its crude oil from Africa compared to 15 percent in 2004. President Bush appeared to have African oil supplies in mind during his 2006 *State of the Union Address*, when he announced his intention "to replace more than 75 percent of (U.S.) oil imports from the Middle East by 2025."[69] Doctor Okumu's speculations regarding intentions aside, it is quite clear that oil exports from Africa are strategically important and growing in importance for the world markets yearly. To highlight this point, China accounted for 40 percent of the world's growth in oil demand for the period 2002-2007.[70] To fulfill its growing demand, China is actively seeking new sources of crude oil from African markets.

Oil is not the only mineral of value which the U.S. requires from Africa. Strategic non-fuel minerals are a major element of resource geopolitics.[71] A common definition of a strategic mineral is a mineral that would be needed to supply the military, industrial, and essential civilian needs of the United States during a national emergency or to maintain a vital industry; and they are not found or produced in the United States in sufficient quantities to meet the need.[72] In the 1960's the U.S. Congress became concerned regarding the USSR's growing interest in Africa. To try and ascertain what the Soviet's interests might be, the Congress dispatched a congressional mission to Africa to determine the availability of strategic minerals on the continent. The team's report referred to the area of South Africa, Democratic Republic of Congo, Zambia and Zimbabwe as "the Persian Gulf of strategic minerals".[73] Maintaining the availability of strategic minerals from Africa has remained a security interest for the U.S. ever since. The number of U.S. strategic non-fuel minerals is over ninety[74] and countries in Africa are the primary suppliers for many of the most critical.[75] The continent ranks first or second worldwide with its concentrations of bauxite, chromium, platinum, diamond, gold, cobalt, and manganese to name a few.[76] Of these, there are four important minerals that the United States imports vast quantities from African nations: chromium, cobalt, manganese, and platinum.[77] To highlight these minerals' importance, a few details will follow.

Perhaps the most important strategic mineral for the U.S. is chromium. Industry in the U.S. uses the mineral to manufacture stainless steel and tool steel. Additionally, it is used in many different military and aerospace industry applications. For over 45 years, the U.S has been totally dependent on chromium imports having no commercially available domestic source of its own. The countries of South Africa and Zimbabwe contain 98 percent of the world's reserves of this mineral.[78] Likewise, since 1971, Cobalt is another mineral which has not been commercially available from domestic sources. This mineral is critical for the production of gas turbines and jet engines. The African nations of Zambia, Morocco, the Democratic Republic of Congo and Botswana control a large portion of the world's known supply.[79] The list of non-domestically available minerals continues with manganese, another mineral used in the production of steel. Approximately 39 percent of the U.S. requirement for manganese comes from South Africa.[80] South Africa also provides the vast majority of the U.S. requirements for platinum.[81] Sales of the platinum metals group, which includes palladium and rhodium, were twice those of gold last year.[82] One of the primary uses of platinum is for auto catalysts.[83] The demand for this metal is growing rapidly on the world market due to an increase in auto emission standards and simply to meet the growing demand for cars in developing nations such as China and India.

Many countries in Africa have expanding economies due to the demand for their mineral resources. Although the resource-rich countries in Africa attract more investment, they are more likely to be politically repressive and thus unstable.[84] This instability contributes to other issues within the country and makes them more susceptible to the affects of climate change and other adversities. Countries involved economically in Africa must be aware of their impact and work cooperatively to improve the governance institutions on the continent and thereby improve the overall resilience of the societies. Countries vying for African resources must be aware of the possible resurgence of "resource nationalism" as in the states of Angola, Nigeria, Sudan, Equatorial Guinea, Congo and Chad[85] and work cooperatively to minimize its development vice taking advantage of it. In the long run, it is better for both suppliers and consumers of the resources to encourage stability and good governance in order to increase the mutual benefits.

China is one of the primary consumers of African resources and a major source of foreign direct investment for the continent. In its quest for a closer strategic partnership with Africa, China has increasingly dynamic economic, political, and diplomatic activities on the continent.[86] China's overall investment in Africa is significant. China's demand for resources and new markets for its manufactured goods makes Africa a natural strategic area of interest. China has signed numerous trade deals with African nations. These agreements usually take the form of loans in exchange for resource concessions. One such deal involved Congo. It agreed to a $5 billion dollar loan from China in exchange for resource concessions and highway access. The country will use the financial boost to rebuild aging infrastructure and upgrade its mines.[87]

Other Chinese investment deals include Mauritanian iron ore (August 2007)[88]; African oil production including buying a significant stake in Sudan's major oil fields and operations in seven other countries; oil exploration agreements in an additional six African countries (Algeria, Angola, Congo, Gabon, Mali and Sudan); and the China National Offshore Oil Corporation is investing $2.3 billion in Nigeria's offshore oil fields.[89] These represent only a portion of the Chinese strategic positions in African resources and the $10 billion in concessional loans to Africa for the period 2006-2008.[90] This trend is likely to continue as long as China has an abundance of cash reserves accumulated from its favorable trade imbalance.

China develops many of these investment arrangements through regular Sino-African Summits. The most recent summit was held in November 2006 in which 48 of the 54 African nations participated.[91] More than one African observer feels China's financial influence "sets the stage for a new balance of power within which Africa will be better able to negotiate with external development partners."[92] The coordination of international investment in Africa, and the inducement for political and social change which might be encouraged with the investment, is a possible area of engagement for AFRICOM which this paper will discuss later.

Why U.S. African Command (AFRICOM)

As indicated earlier, *The National Security Strategy of the United States of America* (March 2006) clearly indicates why Africa is strategically important to the U.S. In brief, the 2006 NSS indicates that, "Africa holds growing geo-strategic importance and is a high priority of this Administration."[93] Africa or African nations are mentioned multiple times within the document due to its relevance regarding transnational terrorism, defusing regional conflicts, ending tyranny, promoting effective democracies, economic growth and free trade, and protecting against the proliferation of weapons of mass destruction – almost every segment of the document includes references to Africa. The message is clear, Africa is no longer an afterthought in U.S. foreign policy. It has expanded strategic importance and the U.S. government (USG) must develop the strategies to address the nation's interests there. Where does AFRICOM come in? The 2006 NSS provides additional guidance –

> The United States recognizes that our security depends upon partnering with Africans to strengthen fragile and failing states and bring ungoverned areas under the control of effective democracies. Overcoming the challenges Africa faces requires partnership, not paternalism. Our strategy is to promote economic development and the expansion of effective, democratic governance so that African states can take the lead in addressing African challenges. Through improved governance, reduced corruption, and market reforms, African nations can lift themselves toward a better future. We are committed to working with African nations to strengthen their domestic capabilities and the regional capacity of the AU to support post-conflict transformations, consolidate democratic transitions, and improve peacekeeping and disaster responses.[94]

Taken by itself, the NSS would not necessarily imply that the Combatant Command, in this case AFRICOM, would have a large role in the areas indicated above (strengthening fragile and failing states, effective democracies, economic development, improved governance, etc.). However, when combined with Department of Defense Directive 3000.05, *Military Support for Stability, Security, Transition, and Reconstruction (SSTR) Operations* (DoD 3000.05), the guidance becomes clearer.

The stated purpose of DoD Directive 3000.05 is quite unambiguous; "establish DoD policy and assign responsibilities within the Department of Defense for planning, training, and preparing to conduct and support stability operations."[95] The document continues by indicating that stability operations are the responsibility of both civilian and military organizations and are conducted throughout the range of full spectrum operations. The directive is explicit in indicating the importance of civil-military cooperation. Cooperation may require working with a wide variety of national and international activities. Department of Defense tasks may include ensuring security, developing local governance structures, promoting bottom-up economic activity, rebuilding infrastructure, and building indigenous capacity for such tasks.[96] What does this mean for AFRICOM? In addition to the DoD tasks indicated above, Directive 3000.05 also specifies additional tasks for the Combatant Commands. Their tasks include designating a Joint Force Coordinating Authority for Stability Operations; incorporating stability operations into military training, exercises, and planning; and engaging relevant USG activities, foreign governments, IOs, NGOs, and members of the private sector in concerns regarding stability operations. These tasks set the stage for a good portion of AFRICOM's mission and subsequent engagement in areas which are being adversely affected by climate change. If AFRICOM can positively impact stability operations on the continent, it will by default improve the underlying governance, security and economic institutions which will allow Africa to face the challenges of climate change.

AFRICOM is unique within the Unified Commands. It will have a unique structure and a unique mission due in large part to the U.S. strategic security interests and the regional requirements of the continent. Theresa Whelan, Deputy Assistant Secretary of Defense for African Affairs, identified these unique characteristics in her testimony before the Senate Foreign Relations Committee in August 2007 when she indicated it would be an "innovative command". She went on to lay out several of the command's most important missions: building African regional security and crisis response capacity; building partnerships and theater security cooperation; improving counter-terrorism skills in African nations; supporting U.S. government agencies in implementing their programs; and always working towards a goal of promoting regional stability. She correctly emphasized that as Africa moves to strengthen its regional security structures, AFRICOM needs to be there to engage on a regional basis to support their efforts.[97]

Partnership, development of African capabilities, and fostering good governance are the keys to this strategy. What is fundamentally different from previous U.S. strategies for Africa is the understanding that the development of good governance is the key to most other issues in Africa. Fostering the institutions which support good governance will significantly improve the rate of success for other programs designed to tackle the multitude of issues the continent faces. As climate change and the adverse impacts it may bring can be both local and regional, AFRICOM's continental perspective will allow it the perspective to effectively partner with African, regional and USG entities. Partnerships will allow it to coordinate different aspects of the response to climate change. The structure of AFRICOM will aid in facilitating the partnership relationships with other USG activities as the Headquarters will include individuals from these same organizations. Agencies represented include the Department of State, the U.S. Agency for International Development (USAID) and others.

The permanently assigned interagency staff at AFRICOM will bring immense capabilities to the headquarters in such areas as coordinating peacekeeping efforts, disaster relief, humanitarian support and aid packages. The diverse skill sets and ties to other USG agencies will allow the headquarters to proactively engage on issues relevant to the continent but at a more local level of expertise. To assist in managing this diverse staff, there will be two deputies, one military and one civilian. The Deputy to the Commander for Civil-Military Affairs (DCMA) will be a senior Foreign Service Officer from the Department of State. This civilian deputy will be responsible for the planning and oversight of the majority of AFRICOM's security assistance work.[98] As the impacts from global climate change stress the African continent's systems more and more, AFRICOM will be well structured to plan, train, coordinate, and react to regional crises in a manner which will support African institutions.

Although *The Army Strategy for the Environment* (2004) was written before the Army outlined AFRICOM's mission, the policy has relevance and application to how AFRICOM can and should impact issues arising from climate change. The Army Environmental Policy Institute's goals and *The Army's Strategy for the Environment* dovetail well with issues AFRICOM will face on the continent. While *The Army Strategy for the Environment* doesn't mention climate change directly, it does highlight several areas which could overlap with AFRICOM's mission. For example, it indicates that regional issues such as natural disasters, environmental damage and famine, coupled with political and social instability, are developing into global issues that will impact the U.S. Military engagement, in cooperation with global, federal or local authorities, may be required on the part of the Army in order to stabilize the situation.[99]

Environmental issues may not top the list of concerns for the U.S. and subsequently AFRICOM. However, a cursory review of African media would indicate this subject is rising rapidly as one of the continent's growing issues. The awareness of the detrimental impact of environmental issues, either brought on by population increases, resource degradation, economic competition, or climate change, is growing exponentially. Media attention and the African nations' governmental pressure will likely highlight environmental and climate change issues to AFRICOM. How AFRICOM responds to the African environmental challenges, and possible violent conflicts, will likely contribute significantly to the overall African reception and attitude towards AFRICOM. Currently, there is work to do concerning the opinions coming from Africa regarding AFRICOM and what the true U.S. plans for the headquarters might be.

The negative perception of AFRICOM and its mission in the eyes of many Africans is certainly a hurdle it will have to overcome before it gains full legitimacy and can perform its mission to the utmost potential. Many Africans, IOs, and NGOs see AFRICOM as militarizing U.S. foreign policy and actually hampering future efforts on the continent. Mark Malan, the Peace Building Program Officer from *Refugees International*, sees an important role for AFRICOM, but strictly in regards to security improvement issues. He highlighted his concerns in his Congressional testimony in August 2007. During this testimony he stressed the need for greater interagency cooperation in regards to Africa. However, he highlighted the concern he and other African experts have of putting non-military issues under the prevue of AFRICOM. He felt this sent the wrong signal; possibly a militarization of the U.S. African policy and further "subjugation and co-option" of U.S. foreign policy. He could envision a liaison between humanitarian, developmental and military activities, but not integration.[100]

Mr. Malan's testimony made several excellent points which highlight the fine line AFRICOM, the Defense Department, and the State Department will have to walk in order to effectively build consensus regarding AFRICOM's mission. Is AFRICOM, as structured, even the right organization to execute its mission set successfully? Without a doubt, the answer is "Yes." The most important thing Africa needs is security. AFRICOM's undivided attention to the African continent will enable unequalled support in establishing the area of security enhancement. Working with African organizations to establish security and good governance must be the first priority for AFRICOM (as these institutions will set the foundation for all other activities). AFRICOM is the appropriate mechanism at the appropriate time to energize the process and effectively tackle not only security issues, but a wide variety of full spectrum operations. The DoD is the only USG activity which has the resources to tackle such an expansive mission set which will include the impacts from climate change. As General Ward, AFRICOM's first commander recently indicated, "AFRICOM will be a learning organization. The AFRICOM that exists today will evolve and will look different in the future as we gain better understanding through our work with others."[101]

How AFRICOM Engagement can Mitigate the Effects of Climate Change in Africa

Climate change is a growing reality which brings the U.S. opportunities as well as challenges on the African continent. To take advantage of these opportunities, mitigate the security threats, and to promote U.S. national security interests there, we must do the following.

- develop regional climate change expertise within the AFRICOM staff (force structure, technical training, and African scientific community engagement)
- focus on dual-use equipment when providing items for African militaries
- implement the stability establishment concepts within DoD 3000.05
- partner with European nations; maximize their African expertise
- engage China on African issues; look for areas to cooperate/moderate
- advocate climate change impacts (sponsoring conferences, engagement within the USG, outreach to the African scientific community)
- emulate the success of other Combatant Commands on environmental issues

Develop Regional Climate Change Expertise

The first, and likely most important, recommendation is to gain awareness of the issue of climate change and its sensitivity within the African populations. This includes awareness from a global perspective in line with *The Army Strategy for the Environment*, the UN's *Intergovernmental Panel on Climate Change* report ("Climate Change 2007"), African scientific and media reports, and other general information on the subject. There is a vast collection of work on the impact of climate change on Africa. Many of the documents are regional or local in nature and will be useful in contingency planning, allocation of resources, and topics for engagement at the national and regional level. In many cases, climate change and the resulting impacts will be an overarching "condition" when examining other issues on the continent and maintaining a level of awareness, if not even expertise, is recommended.

There are many possible options that AFRICOM might use to assist in mitigating the current and future effects of global climate change in Africa. In some instances it will take AFRICOM's direct involvement and leadership on an issue. In some cases, AFRICOM must play a supporting role to other USG agencies. An example of such an instance would include supporting scientific research access. In some instances, not hindering a proposal may be all the support necessary for an IO, NGO or other USG activity to be successful. Actively engaging subject matter experts, maintaining awareness of the subject and aggressively working the issues before they become a crisis are key for AFRICOM's success.

To accomplish the tasks above, correct staffing is essential. To analyze and coordinate issues associated with climate change, the AFRICOM staff must contain the requisite expertise to work the issues. The current proposed staff organization reflects a

Science and Technology Section under the Director of Resources. Among other duties, this section should contain a staff possessing the skills necessary to evaluate the current and long term threat associated with climate change in Africa. The study of climate change is a multi-disciplinary subject, therefore selection of candidates for these positions should take into account experience in the specific areas of concern. The author recommends at least one member of the Science and Technology Section be a qualified climatologist and one be a meteorologist familiar with African weather patterns. Building this staff with the appropriate expertise will provide the AFRICOM Headquarters with invaluable insight and coordination capability regarding the climate change issue.

Dual-Use Equipment

Resource availability may likely determine the level of success in Africa. With this in mind, when nations discuss donor military equipment for Africa, dual-use equipment (all-terrain vehicles, radios, GPS and small transport aircraft that can support climate change related issues such as disaster response, scientific research, and other requirements) should always be a consideration. AFRICOM's primary mission is to assist in building the underlying security and governance structures in Africa. However, in many African countries the military is the only organization which can be organized to respond to a crisis or threat regardless of whether this is a security, environmental, or other.[102] If dual-use equipment is the predominant type provided African nations, then the countries may use it to respond to any type situation to include humanitarian. Currently, only 3 percent of the total United States' $9 billion aid and development package provided Africa is in the form of Military Security Assistance Programs.[103] This indicates that the amount of equipment in question is relatively small (although equipment is also available from other sources to include other countries). The limited amount of equipment in question only emphasizes the benefits by prioritizing not only the U.S. equipment, but also coordinating with other countries on the equipment they are providing to ensure maximum synergy and multi-use applications. Additionally, some of this equipment is not military specific in origin and offers a potential for non-U.S. countries to contribute equipment in support of AFRICOM objectives. An excellent example of this type support is small, all-terrain, Japanese pick-up trucks.

To add specificity to this recommendation, Mark Malan's congressional testimony again provides some keen insight to a way ahead regarding the management and coordination of donated equipment. He feels strongly that AFRICOM can provide a valuable contribution in this area because most African nations are unable to effectively screen and harmonize numerous good intended donor contributions; equipment comes from a wide variety of organizations such as the EU, G8, P3 (U.S., GB, FR), Nordic countries and others. Mr. Malan feels AFRICOM's role as donor coordinator could be an important tool to build African security capacity.[104] With focus from AFRICOM, the command can guide the contributing nations and organizations towards the most efficient and effective combination of equipment designed to meet the dual use demands of the continent.

Implement DoD 3000.05

The next recommendation only warrants a brief reiteration of previous points. The Army establishment in general and AFRICOM in particular must embrace the importance of stability operations as outlined in DoD Directive 3000.05. Early indications are that *U.S. Army Field Manual, FM 3.0, Operations,* (in draft at this paper's writing) provides significant emphasis in this area. Focusing AFRICOM assets in the vital areas associated with the stability of a country's institutions will provide the long term foundation upon which to execute other operations (i.e. counterterrorism).

Partner with Europeans

Several countries currently have greater experience and presence in Africa than does the U.S. These include Great Britain and France. Inviting these countries to provide liaison officers to the AFRICOM Headquarters would benefit each country's efforts. The range of possible issues which could be worked in a direct coordination manner is countless. As GEN Ward indicated, "AFRICOM is a learning headquarters." The mix and number of liaison officers could be adjusted as the command gained experience and determined where it could use the direct inter-country coordination. In regards to climate change, the European Union countries have a well developed appreciation of the potential damage from climate change. Outwardly, their sensitivity to the issue appears greater than that found in the U.S. Therefore, the likelihood of support of plans associated with mitigating the effects of global warming could find cooperative associates in EU countries. Although military liaison officers may not be the correct channel in which to work these issues, other AFRICOM staff should explore the possibility of garnering EU support and cooperation. For example, developing contingency plans for disaster support, cooperatively working security for IO and NGO efforts, and sponsoring climate change mitigation conferences. There is precedence for U.S.-European cooperation on similar issues. Since the early 1990's, NATO has effectively engaged former East Bloc nations using Environmental Security as an inroads to other topics of discussion.[105]

Engage China

In the spirit of inter-military/government engagement, this paper recommends actively seeking to engage Chinese entities in regards to Africa. The rate of economic and political involvement by the Chinese in Africa is growing steadily. A recent study published by the Center for Strategic and Internal Studies (CSIS) indicates that "China's quest to build a strategic partnership with Africa fits squarely within Beijing's global foreign policy strategy and its vision of the evolving international system;" and that "policymakers believe it is in China's interest to engage third parties on Africa, but cautiously, slowly, and with serious reservations."[106] Due to its own strategic concerns, primarily resource access and expanding markets for its consumer goods, Africa is extremely important to China and they will be there for the long term. It is the interest of the Chinese, the U.S. and Africans that Africa continues on a path of peaceful transition to good governance and security. Without these two foundational conditions, both countries may find resource and market availability interrupted and more importantly, Africa will find itself unable to develop the capabilities to mitigate the effects of climate change or other challenges. If this happens, relatively small, short term events (localized flooding for example) will have significant detrimental impacts.

China has signed multiple economic deals with African nations where China will receive resource access in return for loan concessions. There are two views to the China aid story which AFRICOM and the USG need to consider and watch. Stephanie Hanson, writing for the Council on Foreign Relations, advances these two viewpoints in a recent article. One camp sees China's "no-strings-attached" loans as undermining U.S. efforts to improve the good governance practices of developing African nations. The other camp sees China's interest in Africa as a potential benefice and boost to efforts to improve African economic conditions.[107] Claudia E. Anyaso, provides additional insight into U.S. and China relations vis-à-vis Africa with her remarks at a recent *China in Africa Today Seminar*. She reiterated that it is not the U.S. policy to hinder Chinese engagement in Africa. She also encourages cooperation with an eye towards minimizing the impact of Chinese policies which may be out of step with the U.S. She sees positive signs that China is moving towards becoming a "responsible stake holder" in the world. She highlights this point by indicating China's involvement in recent peacekeeping operations in different parts of Africa (Liberia, the Democratic Republic of Congo, and southern Sudan).[108]

AFRICOM's challenge is to encourage the engagement of China and assist in drawing them into the process and thereby being able to work together to coordinate development aid packages and other efforts. Although this appears to be outside the subject matter at hand, mitigating climate change in Africa, the underlying strengthening of basic institutions serves the end goal of improving the resiliency of African institutions. Better prepared institutions will be more able to meet the challenges of climate change.

<u>Advocate Climate Change Impacts in Africa</u>

Although much research has been done regarding climate change in Africa, the need for sustained, long term research is still quite necessary. Although AFRICOM would not conduct the studies necessary to provide more definitive research, the headquarters could act as a coordinating agent in this effort. There are many aspects to this – determining requirements in association with U.S. and African researchers; providing or improving security efforts in research study locations; sponsoring conferences on the subject to raise awareness and determine priorities; requesting and coordinating resources from other USG agencies; advocating research in Africa; and numerous other efforts. In addition to gaining a better understanding of issues in its area of responsibility (AOR), the efforts would also pay dividends in regards to developing regional relationships and support for the headquarters.

Emulate Other Combatant Commands' Environmental Programs

Environmental issues are an area where AFRICOM should build on the success of other more established Combatant Commands. There are several well established programs in this area. The DoD and the Geographic Combatant Commanders use Environmental Security as a means to engage other nations; both their militaries and civil agencies. The relationships fostered and skills taught contribute to the building of good governance and democratic institutions. Other USG agencies often contribute to the DoD efforts. These include the U.S. Agency for International Development (USAID); U.S. Geological Society; Environmental Protection Agency; and Department of the Interior.[109]

A command with documented success in the area of Environmental Security area is the United States Southern Command (SOUTHCOM). Their well established environmental training program attracts a wide range of students and expert instructors. A cornerstone of the program is their Environmental Security Training Workshop. These training sessions provide the military, civil police and environmental officials of Central and South America with the tools and knowledge to return to their countries and teach environmental security issues. Additionally, the training reinforces civil-military cooperation and builds capabilities to address environmental issues that are related to regional security. The SOUTHCOM program has strengthened the military's role in environmental response and has built governmental legitimacy and respect for the armed forces. This type of engagement by AFRICOM is essential as it will help establish the institutional foundations necessary for African nations to effectively support their population's issues as well as support U.S. strategic objectives. To obtain greater relevancy in the region, the leadership of AFRICOM should recognize the importance of environmental issues in their largely humanitarian efforts. Experienced Africa experts concur with the assessment that "environmental security issues determine stability in much of Africa and the effects of climate change will greatly affect this relationship and very likely the engagement strategies of other regional commands."[110] Carrying the perspective forward, that environmental issues are a priority, will serve AFRICOM well.

Conclusion

Three facts are now clear: global climate change is occurring due to an increase in global warming; climate change presents a new and very different type of national security challenge[111]; and unless the U.S. takes proactive engagement steps to mitigate the adverse impacts of climate change on Africa, "climate change will function as a threat multiplier and conditions will develop which are conducive to the development of regional and international security threats".[112] The Departments of Defense and State and the remainder of the interagency have a unique opportunity with AFRICOM. They must gather the best practices developed within the other Combatant Commands,

NGOs, IOs, African organizations, and other activities to make its engagement on the continent truly effective. There is political momentum within the U.S. Congress on climate change. AFRICOM must establish its Environmental Security Program priorities and policies in line with the DoD and the *National Security Strategy* in order to take advantage of this political momentum. There is a brief window of opportunity as AFRICOM establishes itself to make engagement on climate change impacts a top priority of its mission.

Endnotes

[1] Intergovernmental Panel on Climate Change (IPCC), eds., *Summary for Policymakers. In: Climate Change* 2007: *The Physical Science Basis* (New York: Cambridge University Press, 2007), 5. Here after cited as IPCC.

[2] Center for Naval Analysis (CNA), *National Security and the Threat of Climate Change* (Alexandria, VA: Center for Naval Analysis, 2007), 3; available from http://securityandclimate.cna.org/; Internet; accessed 15 September 2007. Here after cited as CNA.

[3] Ian Sample, "Heat – How We Got Here," in *The Reference Shelf: Global Climate Change*, ed. Paul McCaffrey (New York: H.W. Wilson, 2006), 5.

[4] Ibid., 6.

[5] Ibid.

[6] Ibid.

[7] Ibid., 7.

[8] Ibid.

[9] Ibid.

[10] Ibid., 8.

[11] Norwegian Nobel Committee, "THE NOBEL PEACE PRIZE FOR 2007," available from http://nobelpeaceprize.org/eng_lau_announce2007.html; Internet; accessed 14 September 2007.

[12] Ibid.

[13] IPCC, 2.

¹⁴ Ibid.

¹⁵ Ibid.

¹⁶ Ibid., 5.

¹⁷ Dr. Robert M. Carter, "Public Misperceptions on Human-Caused Climate Change: The Role of the Media," Testimony before the Senate Environment and Public Works Committee; 109th Cong., 2nd sess., 6 December 2006; available from http://epw.senate.gov/ hearing_statements.cfm?id=266555; Internet; accessed 4 November 2007.

¹⁸ David Ellsworth, "Researchers Say Humans Can't Rely On Plants to Store Excess CO_2 Forever," *University of Michigan News Service*, 12 April 2006; available from http://www.ns.umich.edu/htdocs/releases/print.php?htdocs/ releases/plainstory. php?id=208&html; Internet; accessed 10 December 2007.

¹⁹ Marlo Lewis, Jr., "Scare Mongering as Journalism: A Commentary on *Time's* "Special Report" on Global Warming," available from http://www.cei.org/pdf/5288.pdf; Internet; accessed 4 November 2007.

²⁰ CNA, 3.

²¹ Dr. Kent Hughes Butts, "The National Security Implications of Climate Change," Testimony before the Subcommittee on Investigations and Oversights, House Committee on Science and Technology; 110th Cong., 1st sess., 27 September 2007, 4; available from http://democrats.science.house.gov/Media/File/Commdocs/hearings/2007/oversight/27sept/butts_testimony.pdf; Internet; accessed 9 October 2007.

²² Ibid.

²³ George W. Bush, *The National Security Strategy of the United States of America* (Washington, D.C.: The White House, March 2006), 24.

²⁴ The CNA Military Advisory Panel members included General Gordon R. Sullivan, USA (Ret) (Panel Chairman), Admiral Frank Bowman, USN (Ret), Lieutenant General Lawrence P. Farrell Jr., USAF (Ret), Vice Admiral Paul G. Gaffney II USN (Ret.), General Paul J. Kern, USA (Ret.), Admiral T. Joseph Lopez, USN (Ret.), Admiral Donald L. "Don" Pilling, USN (Ret.), Admiral Joseph W. Prueher, USN (Ret.), Vice Admiral Richard H. Truly, USN (Ret.), General Charles F. "Chuck" Wald, USAF (Ret.), General Anthony C. "Tony" Zinni, USMC (Ret.).

²⁵ CNA, 3.

²⁶ Ibid.

²⁷ James R. Greenburg, *National Security Policy and Strategy, Course Directive Academic Year 2008* (Carlisle Barracks, PA: U.S. Army War College, 2007), 4.

²⁸ Doug Struck, "Warming will Exacerbate Global Water Conflicts," *Washington Post*, 20 August 2007; available from http://www.washingtonpost.com/wp-dyn/content/article/ 2007/ 08/19/AR2007081900967.html; Internet; accessed 16 October 2007.

[29] General George W. Casey Jr., "The Strength of the Nation," *Army Magazine,* 57 (October 2007); 19; available from http://www.ausa.org/webpub/DeptGreenBook.nsf/byid/WEBP-77GMGC/ $File/Casey.pdf?OpenElement; Internet; accessed 14 October 2007.

[30] Ibid., 20.

[31] Ibid., 21.

[32] CNA, 6.

[33] Butts, 4.

[34] Donald H. Rumsfeld, "The National Security Implications of Climate Change," *Quadrennial Defense Review Report*, (Washington, D.C.: U.S. Department of Defense, 6 February 2006), quoted in Dr. Kent Hughes Butts, Testimony before the Subcommittee on Investigations and Oversights, House Committee on Science and Technology; 110th Cong., 1st sess., 27 September 2007, 4; available from http://democrats.science.house.gov/Media/File/ Commdocs/ hearings/2007/oversight/27sept/butts_testimony.pdf; Internet; accessed 9 October 2007.

[35] U.S. Department of Defense, *Directive 3000.05, Military Support for Stability, Transition, and Reconstruction (SSTR) Operations* (Washington, D.C.: U.S. Department of Defense, November, 2006), quoted in Dr. Kent Hughes Butts, Testimony before the Subcommittee on Investigations and Oversights, House Committee on Science and Technology; "The National Security Implications of Climate Change," 110th Cong., 1st sess., 27 September 2007, 4; available from http://democrats.science.house.gov/Media/File/ Commdocs/ hearings/2007/ oversight/27sept/butts_testimony.pdf; Internet; accessed 9 October 2007.

[36] Charles Cobb Jr., "Africa: Warming World Challenge for Africa," 15 June 2001; available from http://allafrica.com/stories/200106150466.html; Internet; accessed 2 October 2007.

[37] CNA, 6.

[38] Ibid., 20.

[39] George Manful, "Capacity-Building Initiatives to Implement the Climate Change Convention in Africa," in *Climate Change and Africa*, ed. Pak Sum Low (New York: Cambridge University Press, 2005), 297-298.

[40] "Africa: Food Production to Halve By 2020," 25 September 2007; available from http://allafrica.com/stories/200709251134.html; Internet; accessed 2 October 2007.

[41] Cobb, "Africa: Warming World Challenge for Africa."

[42] Struck, "Warming will Exacerbate Global Water Conflicts."

[43] Peter J. Ashton, "Avoiding Conflicts over Africa's Water Resources," *AMBIO: A Journal of Human Environment* (May 2002): 236-242; available from http://ambio.allenpress.com/ perlserv/?request=get-document&doi=10.1639%2F0044-7447%282002%29031%5B0236% 3AACOASW%5D2.0.CO%3B2; Internet; accessed 18 September 2007.

[44] Ibid.

[45] Jonathan A. Patz, et al., "Impact of Regional Climate Change on Human Health," *Nature* (17 November 2005): 310.

[46] Martin Adjuik, et al., "Cause-Specific Mortality Rates in Sub-Saharan Africa and Bangladesh," *Bulletin of the World Health Organization* (March 2006): 185; available from http://www.who.int/bulletin/volumes/84/3/181.pdf; Internet; accessed 10 October 2007.

[47] Cobb, "Africa: Warming World Challenge for Africa."

[48] Struck, "Warming will Exacerbate Global Water Conflicts."

[49] Manful, 298.

[50] U.S. Department of State, *Senator Paul Simon Water for the Poor Act 2005: Report to Congress* (Washington, D.C.: Department of State Publications, June 2007), 29.

[51] World Health Organization, "Sub-Saharan Africa Fact Sheet," available from http://data.unaids.org/pub/EpiReport/2006/20061121_EPI_FS_SSA_en.pdf; Internet; accessed 16 October 2007.

[52] Richard Cincotta, "State of the World 2005 Global Security Brief #2: Youth Bulge, Underemployment Raise Risks of Civil Conflict," 1 March 2005; available from http://www.worldwatch.org/node/76; Internet; accessed 16 October 2007.

[53] CNA, 6.

[54] William R. Cline, *Global Warming and Agriculture Impact Estimates by Country*, July 2007; abstract available from http://allafrica.com/sustainable/resources/view/00011373.pdf; Internet; accessed 17 October 2007.

[55] Dave Opiyo, "Continent Faces Worst Effects From Global Warming," *The Nation*, 13 September 2007 [newspaper available on line]; available from http://allafrica.com/stories/200709130059.html; Internet; accessed 19 October 2007.

[56] Manful, 298.

[57] Cline, *Global Warming and Agriculture Impact Estimates by Country*.

[58] Opiyo, "Continent Faces Worst Effects From Global Warming."

[59] Ibid.

[60] Claudia E. Anyaso, Director, Public Diplomacy and Public Affairs, African Affairs, U.S. Department of State, "China in Africa Today," briefing remarks, The China in Africa Now Seminar, San Antonio, Texas, 6 March 2007.

[61] Cincotta, "State of the World 2005 Global Security Brief #2: Youth Bulge, Underemployment Raise Risks of Civil Conflict."

⁶² Lionel Beehner, "The Battle of the 'Youth Bulge'," Council on Foreign Relations Website; 27 April 2007; available from http://www.cfr.org/publication/13094/ battle_of_the _youth_bulge.html; Internet; accessed 11 November 2007.

⁶³ Stephanie Hansen, "MEND: The Niger Delta's Umbrella Militant Group," Council of Foreign Relations Website; 22 March 2007; available from http://www.cfr.org /publication/ 12920/; Internet; accessed 11 November 2007.

⁶⁴ Manful, 298.

⁶⁵ Manful, 297.

⁶⁶ Greenburg, 4.

⁶⁷ George W. Bush, *The National Security Strategy of the United States of America* (Washington, D.C.: The White House, September 2002), 10.

⁶⁸ Stephen J. Morrison, "Exploring the U.S. Africa Command and a New Strategic Relationship with Africa," Senate Committee on Foreign Relations, Subcommittee on Africa, testimony before the Subcommittee on Africa, 110th Cong., 1st sess., 1 August 2007, 4; available from http://foreign.senate.gov/testimony/2007/MorrisonTestimony070801.pdf; Internet; accessed 15 September 2007.

⁶⁹ Dr. Wafula Okumu, Head of the African Security Analysis Program, Institute for Security Studies, Pretoria, South Africa, "Africa Command: Opportunity for Enhanced Engagement or the Militarization of U.S.-Africa Relations?", House Committee on Foreign Affairs, Subcommittee on Africa and Global Health, testimony before the Subcommittee on Africa and Global Health, 110th Cong., 1st sess., 2 August 2007; available from http://foreignaffairs.house.gov/110 /oku080207.htm; Internet; accessed 15 September 2007.

⁷⁰ Esther Pan, "China, Africa, and Oil," Council on Foreign Relations Website; PODCAST; 3 May 2006; available from http://www.cfr.org/publication/10586/china_africa_and_oil.html; Internet; accessed 5 November 2007.

⁷¹ Ewan W. Anderson and Liam D. Anderson, *Strategic Minerals: Resource Geopolitics and Global Geo-Economics* (New York: Wiley, John and Sons, 1998), on-line abstract; available from http://www.wiley.com /WileyCDA/WileyTitle/productCd-0471974021.html; Internet; accessed 20 October 2007.

⁷² Natalie Johnson, "Strategic Minerals of the United States," 19 November 2002, available from http://www.emporia.edu/earthsci/amber/go336/natalie/newindex.htm; Internet; accessed 21 October 2007.

⁷³ Mabasa Sasa, "The Persian Gulf of Strategic Minerals of Our Earth," *New African* 465 (August – September 2007): 132 [database on-line]; available from ProQuest; accessed 22 OCT 2007.

⁷⁴ G. J. Mangone, *American Strategic Minerals* (New York: Crane Russak & Company, 1984), 30; quoted in Natalie Johnson, "Strategic Minerals of the United States," 19 November

2002, available from http://www.emporia.edu/earthsci/amber/go336/natalie/newindex.htm; Internet; accessed 21 October 2007.

[75] Johnson, "Strategic Minerals of the United States."

[76] George J. Coakley and Philip M. Mobbs, "The Mineral Industries of Africa," 1999; available from http://minerals.usgs.gov/minerals/pubs/country/1999/africa99.pdf; as quoted in Natalie Johnson, "Strategic Minerals of the United States," 19 November 2002, available from http://www.emporia.edu/earthsci/amber/go336/natalie/newindex.htm; Internet; accessed 21 October 2007.

[77] Johnson, "Strategic Minerals of the United States."

[78] Mangone, *American Strategic Minerals*.

[79] Ibid.

[80] R.A. Hagerman, *U.S. Reliance on Africa for Strategic Minerals*, Written Communications Project (Quantico, VA: Marine Corps Command and Staff College, April, 1984); available from http://www.globalsecurity.org/military/library/report/1984/HRA.htm; quoted in Natalie Johnson, "Strategic Minerals of the United States," 20 November 2002, available from http://www.emporia.edu/earthsci/amber/go336/natalie/newindex.htm; Internet; accessed 21 October 2007.

[81] Ibid.

[82] "The New Gold," *Economist* (10 May 2007); available from http://www.economist.com/finance/displaystory.cfm?story_id=9154250; accessed 5 November 2007.

[83] Ibid.

[84] "African Free-for-All," *Economist* (30 August 2007); available from http://www.economist.com/agenda/displaystory.cfm?story_id=9725453; accessed 5 November 2007.

[85] Ibid.

[86] Bates Gill, Chin-Hao Huang and J. Stephen Morrison, *China's Expanding Role in Africa: Implications for the United States*, (Washington, D.C.: Center for Strategic and International Studies, 5 February 2007), synopsis; available from http://www.csis.org/component/option,com_csis_pubs/task,view/id,3714/; Internet; accessed 25 October 2007.

[87] "The New Gold."

[88] "China's Company to Secure Long-Term Iron Ore Supply from Mauritania," 15 August 2007, linked from *The Arab Steel Homepage;* available from http://www.arabsteel.info/total/Long_News_Total_e.asp?ID=387; Internet; accessed 11 October 2007.

[89] "Current Issues: Oil," linked from *Atlas on Regional Integration in West Africa Homepage*, available from http://www.atlas-ouestafrique.org/spip.php?article98; Internet; accessed 5 November 2007.

[90] "The Poor Brothers," linked from *Atlas on Regional Integration in West Africa Homepage*, available from http://www.atlas-ouestafrique.org/spip.php?article94; Internet; accessed 5 November 2007.

[91] Ibid.

[92] Ibid.

[93] Bush, *The National Security Strategy of the United States of America (2006)*, 37.

[94] Ibid.

[95] U.S. Department of Defense, *Military Support for Stability, Security, Transition, and Reconstruction (SSTR) Operations*, Department of Defense Directive 3000.05 (Washington, D.C.: U.S. Department of Defense, 28 November 2005), 1.

[96] Ibid., 3.

[97] Theresa Whelan, "Exploring the U.S. Africa Command and a New Strategic Relationship with Africa," Senate Committee on Foreign Relations, Subcommittee on Africa, testimony before the Subcommittee on Africa, 110th Cong., 1st sess., 1 August 2007, 4; available from http://foreign.senate.gov/testimony/2007/WhelanTestimony070801.pdf; Internet; accessed 15 September 2007.

[98] Ibid.

[99] GEN Peter J. Schoomaker and HON R.L. Brownlee, *The Army Strategy for the Environment* (Washington, D.C.: 1 October 2004), 1.

[100] Mark Malan, "Exploring the U.S. Africa Command and a New Strategic Relationship with Africa", Senate Committee on Foreign Relations, Subcommittee on Africa, testimony before the Subcommittee on Africa, 110th Cong., 1st sess., 1 August 2007, 4; available from http://foreign.senate.gov/testimony/2007/MalanTestimony070801.pdf; 3, Internet; accessed 5 September 2007.

[101] General William E. Ward, remarks made at the *Foreign Press Roundtable*, 3 October 2007 (Washington D.C.) available from http://www.africom.mil/TranscriptWard DiscussesGoals100307.asp; Internet; Accessed 6 November 2007.

[102] Butts, 9.

[103] Pat Mackin, Editorial Comments, *Baltimore Sun*; 12 October 2007; Available from http://www.baltimoresun.com/news/opinion/letters/bal-ed.le.12ooct12,0,7832359.story; Internet; Accessed 9 November 2007.

[104] Malan, 7.

[105] Butts, 3.

[106] Gill, Huang, Morrison, abstract.

[107] Stephanie Hanson, "Vying for West Africa's Oil," 7 May 2007; available from http://www.cfr.org/publication/13281/vying_for_west_africas_oil.html; Internet; accessed 9 November 2007.

[108] Anyaso, "China in Africa Today."

[109] Butts, 3.

[110] Dr. Kent Hughes Butts, United States Army War College, telephone interview by author, 13 March 08.

[111] CNA, Letter of Transmittal.

[112] Ibid., 6.

www.ingramcontent.com/pod-product-compliance
Lightning Source LLC
Chambersburg PA
CBHW081758170526
45167CB00008B/3237